NISTIR 7525

Standards Development for Gas Phase Air Cleaning Equipment in Buildings

Andrew Persily
Cynthia Howard-Reed
Stephanie Watson
Nicos Martys
Edward Garboczi
Heather Davis
Building and Fire Research Laboratory

Prepared for:
U.S. Department of Homeland Security
Science & Technology Directorate
Washington, DC

October 2008

U.S. Department of Commerce
Carlos M. Gutierrez, *Secretary*

National Institute of Standards and Technology
Patrick Gallagher, *Acting Director*

Abstract
Concerns about building security have resulted in increasing interest in gas phase air cleaning (GPAC) and the need for standard methods to determine the effectiveness of these systems. Similarly, the ability to predict their installed performance, based on such standard test data, is becoming increasingly important. The development and application of these standards and prediction tools will provide better protection of building occupants against chemical agents and improve the ability of designers and building owners to identify and specify air-cleaning equipment with a realistic expectation as to installed performance. The National Institute of Standards and Technology (NIST) has conducted an effort to facilitate the development of these standards and predictive tools under a project funded by the Department of Homeland Security. This report describes the following tasks that were carried out as part of this project: an evaluation of gas phase air cleaning technology; an assessment of existing and proposed standards, as well as relevant guidance documents; design of a laboratory-scale microreactor to evaluate media consistent with current and proposed industry approaches; micro-scale modeling to understand the interactions between the media and gaseous contaminants and, building-scale modeling to understand the impacts of air cleaning systems in controlling occupant exposure.

Keywords: air cleaning, building security, filtration, gas phase, standards

Table of Contents

INTRODUCTION 1
GPAC TECHNOLOGY 1
GPAC STANDARDS 3
Existing Standards 3
Standards Under Development 4
Other Relevant Standards and Guidelines 5
MICROREACTOR 7
Design of Benchtop Microreactor 7
Design of Experiments 11
MICRO-SCALE MODELING 12
Transport Model 12
Filter Characterization 12
Model Test Case 13
Modeling Contaminant Transport 15
Links to Measurement of Concentration over Time 17
Future Research Needs 18
BUILDING SCALE MODELING 18
Simulation Results 20
DISCUSSION 24
ACKNOWLEDGEMENTS 24
REFERENCES 24

INTRODUCTION

As concerns about building security increase, the application of particle filtration and gas phase air cleaning has been receiving increasing attention (NIOSH 2003). Unlike particle filtration, there is currently a lack of standards for determining the effectiveness of gaseous air cleaning systems and tools for predicting their installed performance. This lack of standards makes it difficult for designers, building owners and operators, and equipment manufacturers to specify gaseous air cleaning systems and to incorporate them into system designs. To provide protection against gaseous agents to which a building might be exposed, standards and application guidance for gas phase air cleaning are needed.

NIST has initiated a project to advance both the development of the standards and guidance for gas phase air cleaning, and to better understand the detailed performance of these systems to support standards development and to better predict their installed performance. This report presents the findings of several tasks under this project: assessing existing and proposed standards, as well as relevant guidance documents; designing a laboratory-scale microreactor to evaluate media consistent with current and proposed industry approaches; conducting computational fluid flow simulations of the interactions between adsorbent media and gaseous contaminants to better understand the mass transport processes occurring in air cleaning systems; and, performing building-scale modeling to understand the impacts of air cleaning systems in controlling occupant exposure.

GPAC TECHNOLOGY

Currently available non-industrial gas phase air cleaning (GPAC) devices use one or more of the following removal processes: adsorption, chemisorption, ozone oxidation, air ionization, and ultra-violet photocatalytic oxidation. Although several technologies are capable of removing gas-phase contaminants, there is no single device available today that can remove all types of hazardous gaseous contaminants from an indoor environment. As a result, the different types of gas-phase air cleaners can be described by their specific capabilities and limitations.

Adsorption is the most widely used air cleaning mechanism of gas-phase contaminants in non-industrial buildings (VanOsdell et al. 1996) and therefore the focus of this project. Physical adsorption is an exothermic process where organic vapor molecules are attracted to the surface of an adsorbent material and trapped in the material's pores (< 2 nm diameter). Polarity and pore size dictate a material's affinity for specific contaminants. Adsorbents are typically packed in a bed configuration for contaminant removal, but they may also be embedded in fabric filters or coated on surfaces of particulate filters. In addition to media characteristics and contaminant properties, the effectiveness of sorbent air cleaners also depends on the airflow velocity through the media and environmental conditions such as relative humidity and temperature.

There are many different adsorbent materials available; however, activated carbon is the most commonly used adsorbent in heating, ventilation and air-conditioning (HVAC) systems (Henschel 1998). Activated carbon consists of a carbonaceous material (e.g., wood, coal, bark, coconut shells) that has been partially oxidized to create pore sizes ranging from 0.5 nm to 50 nm and surface areas in the range of 1000 m^2/g (NIOSH 2003). Contaminants most likely to be removed by adsorption onto activated carbon tend to be non-polar and have low vapor pressures (< 1.5 kPa). Activated carbon is less effective for volatile, low-molecular weight gases, such as formaldehyde, ammonia, and vinyl chloride.

Other sorbent materials include silica gel, activated alumina, and zeolites. Silica gel and alumina have less surface area than activated carbon (150 m^2/g to 750 m^2/g) but are often used to trap polar compounds, such as formaldehyde and sulfur-based contaminants (NIOSH 2003). Their affinity for water, however, limits their effectiveness under higher humidity conditions. Zeolites are naturally occurring aluminosilicate minerals that are also hydrophilic but tend to have smaller pores (0.3 nm to 3.0 nm), making them more suitable for low molecular weight halides, such as chlorinated fluorocarbons (NIOSH 2003). It is possible to manufacture synthetic zeolites with larger pore sizes and less sensitivity to humidity.

In general, the addition of sorption media to an existing HVAC system is relatively straightforward. However, sorption-based air cleaning systems are associated with a significant pressure drop, which can increase fan energy requirements in HVAC systems and therefore both first and operating costs. Most sorbent media can be regenerated with high temperatures or solvent extraction, but this process can be typically difficult to perform in situ. However, if there are no space or structural limitations, pressure swing adsorption could be used to regenerate installed media. Pressure swing adsorption uses high temperature air at a high pressure to purge contaminants from activated carbon and exhaust them to the outdoors or to a separator. Sorption has another advantage of producing no harmful byproducts. However, there is potential for media to desorb a portion of the originally sorbed species when the air cleaner becomes saturated and/or the upstream contaminant concentration goes to zero. As a result, gas-phase air cleaner performance should be based on both the initial contaminant removal and breakthrough time (Chen et al. 2005). Another concern is the disposal of used media.

Chemisorption can extend the capabilities of adsorption through use of a reactive chemical coating on the sorbent surface that causes certain contaminants to bond to the media. For example, a number of high-vapor pressure chemicals (e.g., hydrogen cyanide and formaldehyde) are not retained on activated carbon by physical adsorption due to their high volatility. Impregnating the carbon with a reactant will bond the contaminant to the substrate and convert it to more benign chemicals. For example, potassium permanganate is used to oxidize formaldehyde into water and CO_2. In addition to potassium permanganate, chemisorption compounds include sodium sulfide, bromine, phosphoric acid, sodium carbonate, and metal oxides. The US military has developed an impregnated carbon, ASZM-TEDA, to specifically remove a wide range of chemical warfare agents. ASZM-TEDA is a coal-based activated carbon impregnated with copper, zinc, silver, molybdenum and triethylenediamine (NIOSH 2003).

While chemisorption has the advantage of removing a wider range of gas-phase contaminants than physical adsorption alone, it is a slower process and is not reversible. Thus, the chemisorption media is not reusable and must be replaced, emphasizing the importance of filter maintenance schedules and proper disposal of spent media. Another drawback to oxidative chemisorption is the potential production of harmful byproducts resulting from chemical reactions. For example, hydrochloric acid may be produced when chlorinated hydrocarbons are oxidized (Godish, 1989).

While the most common, physical adsorption and chemisorption are not the only technologies available for removing gas-phase air contaminants from indoor air. For example, ozone generators and air ionizers have been shown to remove aliphatic and aromatic alkenes (Boeniger 1995; Daniels 2002), but have potential drawbacks related to ozone exposure and potentially harmful byproducts (Weschler 2000; Boeniger 1995). Ultraviolet-photocatalytic oxidation (UV-PCO) is a more recently developed technology that has been shown to remove several classes of volatile organic compounds (VOCs) and can be characterized by lower maintenance requirements and longer service life (Jacoby et al. 1996; Alberici and Jardim 1997; Blake 1994). However, questions remain regarding catalyst poisoning, potentially harmful byproducts and building conditions that impact performance.

So far, questions about many of the non-adsorption-based air cleaners have hindered wide acceptance in the marketplace. As a result, discussions of gas-phase air cleaning equipment standards, test methods and guidance documents have focused on physical adsorption and chemisorption air cleaner technologies.

GPAC STANDARDS

A primary goal of this project is to support the development of the standards needed to effectively implement gas phase air cleaning as a means of improving building safety and security. A first step in achieving this goal is to assess current standards that are relevant to this technology and which may serve as a vehicle to meet the project goals. At this time, there is a notable lack of standards for determining the effectiveness of gaseous air cleaning systems and their installed performance. In addition to standards, there is a need for better guidance on installation, operation and maintenance of these systems. Note that this need for guidance also applies to particulate filtration.

Existing Standards

This section reviews a number of existing standards that apply to air cleaning systems and discusses their relevance to this project and the application of GPAC to building security.

ASME N510-1989, Testing of Nuclear Air Treatment Systems: This standard provides a basis for the development of test programs and detailed acceptance and surveillance test procedures for high efficiency air-cleaning systems for nuclear power plants. It also specifies minimum requirements for the reporting of test results. The standard covers requirements for the post-delivery field-testing of high efficiency air-cleaning systems for both gas phase and particle matter for nuclear power plants and other nuclear applications. The tests covered by this standard are of two types: (a) acceptance tests, which verify that the systems have been correctly installed and meet the requirements of project specifications; (b) surveillance tests, which monitor the condition of the systems. The actual tests described by the standard include the following: visual inspection, duct and housing leakage, structural capability, mounting frame pressure leakage, airflow capacity and distribution, air-aerosol mixing uniformity, high efficiency particulate air (HEPA) filter in-place leakage, adsorber bank in-place leakage, duct damper bypass, system bypass, air heater performance, and laboratory testing of adsorbent. Only the last test relates directly to the removal efficiency of the filter, in this case only for radioiodine removal using ASTM standard D 3803.

ASME N509-2002, Nuclear Power Plant Air-Cleaning Units and Components: This standard identifies and establishes requirements for filters, adsorbers, moisture separators, air heaters, filter housings, dampers, valves, fans, ducts, and other components of nuclear air-treatment systems for a specific application in a nuclear power plant. Requirements for operability, maintainability and testability of systems necessary for the maintenance of system reliability for the design conditions are also included. The standard specifies acceptance testing, including minimum acceptance requirements, in accordance with Standard ASME N510.

The standard covers requirements for the pre-delivery design, construction, and qualification and acceptance testing of the air-cleaning units and components that make up high efficiency air and gas cleaning systems used in nuclear power plants. Qualification and acceptance testing provisions are specified to verify the adequacy of the air-cleaning unit and component design, to verify that components have been properly fabricated and installed, and that the system will perform in accordance with specification requirements. The categories covered by the standard are as follows: functional design, components, packaging, shipping, receiving, storage, and handling of components, installation and erection, quality assurance, and acceptance testing. This standard does not address determination of removal efficiencies.

<u>ASHRAE Standard 62.1-2007, Ventilation and Acceptable Indoor Air Quality</u>: This standard contains various requirements related to both ventilation and indoor air quality, primarily minimum ventilation requirements and various system requirements related to moisture control and other issues. It does not contain requirements for gaseous air cleaning (except in the case of elevated outdoor ozone levels), but its relevancy to this discussion is based on the inclusion of a performance path for compliance referred to as the Indoor Air Quality (IAQ) Procedure. The IAQ Procedure is an alternative to the prescriptive Ventilation Rate Procedure, which specifies outdoor air requirements as a function of space type. The alternate procedure is based on the identification of contaminants likely to be present in the indoor environment and the determination of the levels of ventilation and air treatment (including gaseous air cleaning) required to maintain their concentrations at acceptable levels. Therefore, gaseous air cleaning is important in that it provides a means to comply with this performance procedure, and the development of test methods for evaluating gaseous air cleaning is key to its application.

Standards Under Development

This section discusses the activities of ASHRAE committee SPC 145P, which is working on three test methods for gas phase air cleaning equipment. The first, referred to as 145.1, is a small-scale laboratory method for evaluating media. The second is a laboratory method for evaluating full-scale devices in ductwork and is referred to as 145.2. The last method, 145.3, is a field test for installed performance.

While not yet approved, ASHRAE Standard 145.1P, Laboratory Test Method for Assessing the Performance of Gas-Phase Air Cleaning Systems: Loose Granular Media, is expected to define a small-scale laboratory test method for measuring the contaminant removal efficiency of loose granular sorptive media. Tests are to be performed at elevated gas challenge concentrations under steady-state conditions for comparing the different media, rather than for predicting their installed performance. The standard will also define methods of calculating and reporting results. The approved scope describes the media types that the standard does not apply to as follows: bonded carbon panels, beaded activated carbon, carbon cloths, absorbant loaded nonwovens, dry process carbon composites, and particulate removal equipment.

The testing apparatus and material will only be applicable to granular (pelletized) media with mean particle diameters of less than 5 mm. Two groups of chemical contaminants are expected to be prescribed for use in the testing, outdoor air acid gases and indoor air volatile organic compounds (VOCs), and the user may select one or both of the groups for testing. Although other contaminant groups may be tested, they may not necessarily be used in rating comparisons. There will be requirements for column length and inside diameter for the test canister, which will be included to average out anomalies due to filter media variations and to avoid significant edge/wall effects. The tests are run to a 50 % breakthrough of the initial challenge concentration, where breakthrough is defined as the point at which the downstream contaminant concentration is measurable and begins to rise rapidly.

The second standard being developed by the ASHRAE committee is referred to as 145.2P, with the tentative title Method of Testing Gaseous Contaminant Air Cleaning Devices for Removal Efficiency. This test method will describe how to perform an in-duct, large-scale, sorptive media laboratory test for measuring the performance of gaseous contaminant air-cleaning devices with operating ranges from 0.24 m^3/s to 0.94 m^3/s (500 cfm to 2000 cfm). The standard will include a large-scale test procedure with quality control constraints to measure percent removal efficiency and removal capacity of sorption-based gas phase air cleaning devices, when they are challenged under steady-state conditions by specified gaseous contaminants. The test is intended to simulate the capture performance of commercially available HVAC media under controlled, but representative conditions.

This procedure will incorporate a number of steps that are designed to reduce personal inhalation exposures during testing to levels well below the respective Permissible Exposure Limits (PELs) and should be followed carefully. A primary focus is careful sealing of apparatus leaks, and applications of tests to demonstrate that leaks into the operator work area do not constitute a significant health hazard. The laboratory test apparatus, equipment, test protocol, quality control guidelines, and equipment calibration recommendations are designed to ensure repeatability within ± 10 % of the measured value.

Other Relevant Standards and Guidelines
This section describes other standards and guidelines that are relevant to the goals of this project.

ASHRAE Standard 52.2-2007, Method of Testing General Ventilation Air-Cleaning Devices for Removal Efficiency by Particle Size: This standard establishes a test procedure for evaluating the performance of air-cleaning devices as a function of particle size and is a good example of the type of method of test needed for the gaseous air-cleaning systems. The standard addresses two air cleaner performance characteristics of importance to users: the ability of the device to remove particles from the airstream and its resistance to airflow. When air cleaners are tested and reported for efficiency in accordance with this standard, there is a basis for comparison and selection for specific applications.

This standard outlines a method of loading the air cleaner with synthetic dust to simulate field conditions. The standard defines procedures for generating the aerosols required for conducting the test and provides a method for counting airborne particles of particular sizes in order to calculate removal efficiency by particle size. This standard also establishes performance specifications for the equipment required to conduct the tests, defines methods of calculating and reporting the results obtained from the test data, and provides a minimum efficiency reporting system that can be applied to air-cleaning devices covered by this standard.

IEST: The Institute of Environmental Sciences and Technology (IEST) has prepared a Handbook of Recommended Practices (RPs) Relating to Air Filtration in Cleanrooms and Other Controlled Environments. These documents address filter and filter media testing, filter construction, and various applications of filters in cleanrooms for the removal of airborne particulate contamination. The handbook includes:

   IEST-RP-CC001 (1993): HEPA and ULPA[*] Filters

   IEST-RP-CC006 (2004): Testing Cleanrooms

   IEST-RP-CC007 (1992): Testing ULPA Filters

   IEST-RP-CC0021 (2005): Testing HEPA and ULPA Filter Media

   IEST-RD-CC011 (1995): A Glossary of Terms and Definitions Relating to Contamination Control

   IEST-RP-CC034.1 (2005): HEPA and ULPA Filter Leak Tests

The recommendations outlined in these documents for particulate removal are models of the type of guidance documents for establishing standards and recommended practices for filter testing and media characterization needed for gaseous air-cleaning devices.

NAFA (National Air Filtration Association), Installation, Operation and Maintenance of Air Filtration Systems (1997): This manual was created to provide guidance on the installation, operation and maintenance of air filtration systems and is based on situations encountered during actual field experience and applications. Its prime concern is to help ensure that filters are properly designed, installed, operated, and maintained for maximum efficiency and safety. The manual instructs the user on the following subjects: air filtration framing systems, filter installation, and filter replacement for different type of particle filters and gas phase air cleaning systems.

NAFA Guide to Air Filtration (2001): This document is an information manual intended for those interested in the fundamentals of air filtration, and is required reading to become NAFA Certified Air Filtration Specialists. The manual covers the following topics: the importance of air filtration, principles of air filtration, types of filters, electronic air cleaners, HVAC filter testing, HEPA and ULPA filter testing, controlled environments, airborne microorganisms, gaseous contaminants, indoor air quality, owning and operating costs, and brief descriptions of the ASHRAE Standards 52.1 and 52.2.

---

[*] ULPA refers to ultra low particulate air.

MICROREACTOR

As part of this project, NIST has defined experimental and simulation efforts to better understand gas phase air cleaner performance. One aspect of the experimental program involved the design of a laboratory-scale microreactor for measuring media performance. While the microreactor was not constructed as part of this project, the design work is valuable for use in future projects. This section describes the microreator design and its potential application.

The microreactor was designed for small-scale laboratory measurements of contaminant removal efficiency of loose granular sorptive media. The testing envisioned would examine the sorptive media used in gas-phase air cleaning equipment as installed in an airstream and subsequently challenged with test gases at concentrations that are elevated relative to typical applications. The apparatus design is based on the proposed ASHRAE Standard 145.1, and was being considered for potential use in round-robin tests of ASHRAE 145.1 after it is published. The apparatus was also designed to measure permeability of the sorptive media by measuring the change in challenge gas concentrations/volumes after sorptive media introduction.

As noted above, ASHRAE 145.1 is expected to outline a test procedure with quality control restraints to develop a composite rating of small media samples, in granular or pellet form, when it is challenged under steady-state conditions by a number of gaseous contaminants. Such media include untreated or chemically impregnated activated carbon, activated alumina, and other adsorbent materials or catalysts. The test is to be conducted at elevated gas challenge concentrations, relative to ventilation applications, and at a single temperature and relative humidity (RH), to compare the media rather than directly predict performance in any particular application. Gas analyzers are used to measure the changes in challenge gas concentrations over the course of the experiment. The composite rating for small media is based on percent removal efficiency and removal capacity. Thus, the results of these tests can provide information for the design and selection of air cleaning equipment and the design of air cleaning systems for controlling indoor concentrations of gaseous air contaminants.

Design of Benchtop Microreactor System

The test apparatus was designed with certain modifications of the system in the proposed ASHRAE 145.1P. In particular, these modifications include changes to the media column or reactor, the range in challenge gas concentration, and the experimental range of temperature and RH. In addition, the method of gas analysis was specified as gas chromatography-mass spectrometry (GC-MS) to enable exact identification of the challenge gas and possible secondary reaction products, and a mode to re-circulate the challenge gas over the media bed was added to more closely mimic air recirculation in building ventilation systems. These modifications were selected to enable the benchtop microreactor system to more accurately characterize the air media interaction with various challenge gases under more realistic variables, i.e. temperature, RH, and re-circulation mode. Additionally, the air media itself would be easily analyzed spectroscopically for chemical changes after exposure to the challenge gas.

A schematic of the benchtop microreactor system is shown in Figure 1. Two separate gas cylinders contain the challenge gas/air mixture and an air purge. Gas flow is controlled by mass flow controllers. The RH generator, mass flow controllers, media column, and re-circulating pump are housed in a temperature controlled chamber. The chamber is commercially available and would be modified in-house to maintain constant temperature conditions. The control of RH for the microreactor system is performed using a humidity generator designed for the Simulated Photodegradation via High Energy Radiant Exposure (SPHERE) in the Polymeric Materials

Group of NIST as shown in Figure 2. In this humidity generator, RH is controlled by the controlled mixing of moisture saturated and dry air streams. Air is passed through an air drier and split into two streams. One stream is moistened by routing it through a saturated cotton filter wick in a commercial stainless steel water tank (Figure 3) partly filled with distilled, de-ionized water. The water level in the stainless steel tank is maintained by an automated feedback controlled filling system. The flow rates and proportions of the two streams are controlled by a proportional integral differential (PID) control loop, which is driven by voltages generated by a microprocessor board in response to the difference between the set-point and the measured RH. On leaving the water tank, the saturated air stream is immediately mixed with the dry air stream. The combined stream is then passed to a rotary valve and mass flow controller before passing through the media column.

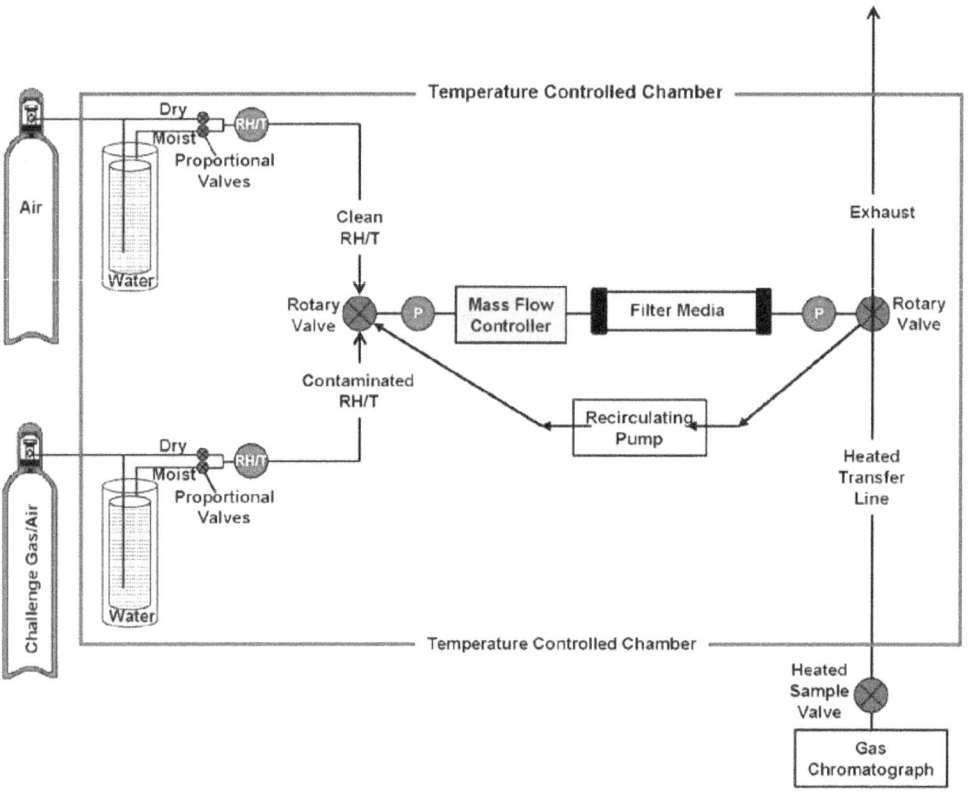

Figure 1. Schematic of the benchtop microreactor system. Green circles with *RH/T* represent relative humidity/temperature sensors. Green circles with *P* represent gas pressure sensors.

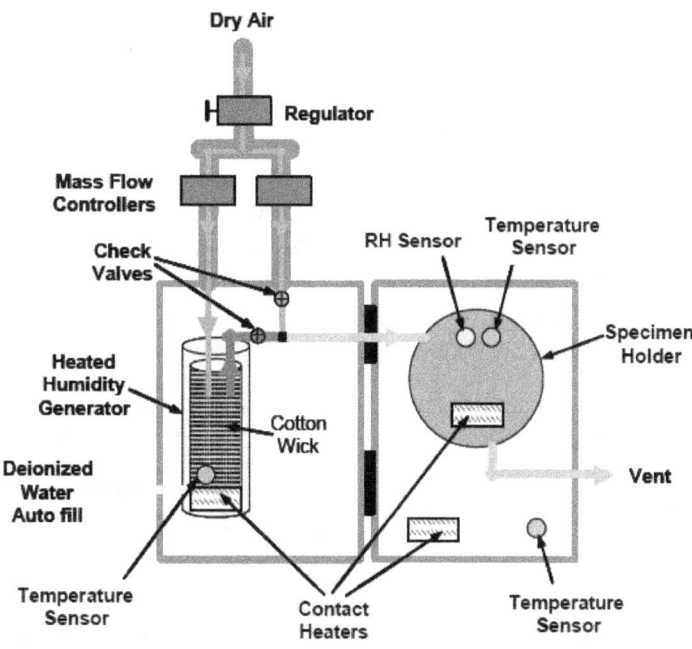

Figure 2. Schematic of the controller for the heating and humidity generation system.

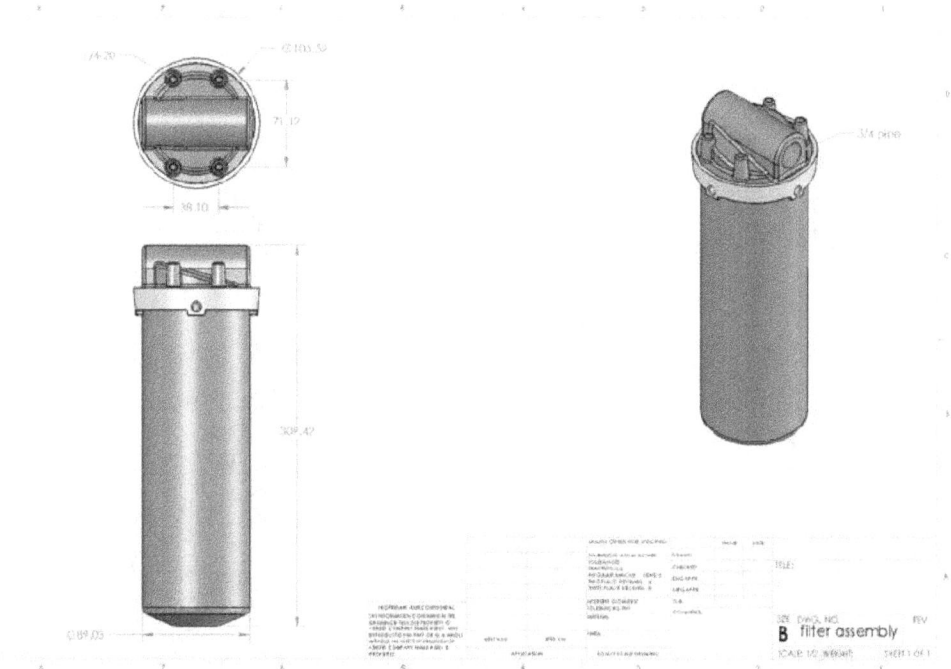

Figure 3. Diagram of a stainless steel water tank used in the humidity generation system. The units for the values of the dimensions are millimeters.

The benchtop microreactor system will accommodate a re-circulating mode using a varistaltic type pump. A means of maintaining RH for the media column system under re-circulating conditions has yet to be determined. However, for most air media samples the addition of water vapor would be needed to maintain a specified level of RH. Moisture may be added by mixing in moist air and/or moist challenge gas/air mixture into the re-circulation loop. In either case, the challenge gas concentration in the re-circulating loop would change. Furthermore, monitoring the RH within the re-circulation loop would require additional RH/temperature sensors to be placed within the re-circulation loop. Initial challenge gas experiments would be conducted without adjusting RH during re-circulation to determine the operating conditions for each air media sample and the amount of additional RH needed to maintain constant RH. RH adjustments would be added at a later phase in the testing.

The media column is modified to insure proper media packing, media bed placement, and adequate seals to maintain pressure and humidity in the system. This media column is constructed mostly of glass, a highly non-reactive material, to reduce air media contamination errors. Glass frits are used as support disks and an inner glass ring, available in adjustable sizes, is used to properly pack the media bed to prevent bed fluidization. The design is outlined in Figure 4.

The analysis method for the contaminant gas and secondary products for this benchtop microreactor is gas chromatography-mass spectrometry with a sample pre-concentrator, for detection of volatile chemicals at low initial challenge gas concentrations on the order of 1 µg/L or less. A commercial rotary valve is used to collect samples of the test gas stream for GC/MS analysis. The mass spectrometer detector allows the direct identification of the challenge gas and secondary products in addition to their concentrations.

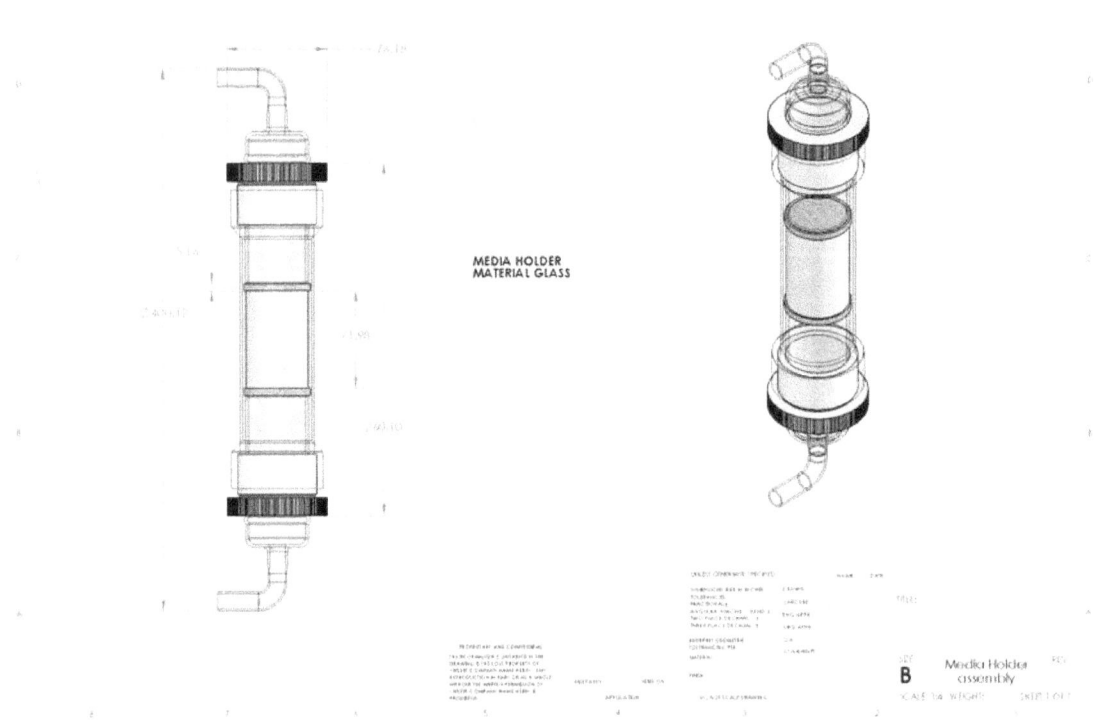

Figure 4. Schematic of the media column. The units for the values of the dimensions are millimeters.

Design of Experiments

A variety of experimental conditions to be used with the benchtop microreactor system are outlined in Table 1. These experimental values were chosen to cover a range of conditions found in buildings and suggested in ASHRAE 145.1P. The allowable variance in the target temperature and target RH, based on ASHRAE 145.1P, is ± 2 °C and ± 5 % RH, respectively. The test airflow rate to be used was that necessary to achieve a 0.1 s residence time, which is theoretical time that an increment of air is within the confines of a media bed, as determined in ASHRAE 145.1P. Residence times and face velocities are calculated using the following relationships:

$$t_r = d/v \quad (1)$$
$$v = Q/A \quad (2)$$

where t_r is residence time in s, d is the media depth in mm and v is the air/gas velocity in mm/s, Q is the air/gas volumetric flow rate in mm^3/s and A is sample face area, equal to πd^2, in mm^2. Given a typical media depth of 25 mm, Q is equal to 0.00049 m^3/s. The variances in the conditions within the benchtop microreactor system are listed in Table 2.

Table 1. Sample experimental conditions for microreactor system challenge gas experiments.

Experiment	Temperature (°C)	Relative Humidity (%)
1	23	50
2	43	50
3	0	50
4	32	90
5	27	65

Table 2. Experimental condition variances allowable within the benchtop microreactor system.

Experimental Variable	Variance
Temperature	± 0.2 °C
RH	± 2 % of value
Airflow	± 1 %
Challenge Gas Concentration	± 0.2 %

The challenge gases to be examined are based on the ASHRAE Standard 145.1P and are categorized into three groups: Acid Challenge Gases, Volatile Organic Carbon (VOC) Gases, and Other Common Challenge Gases. Table 3 outlines the three challenge gas groups and the concentrations that could be used in the experiments. The initial experiments for such studies could focus on VOC challenge gases and examine the top three VOC challenge gases, individually, to obtain preliminary results.

Table 3. Challenge Gases and Experimental Concentrations

Acid Challenge Gases		VOC Challenge Gases		Other Common Gases	
Compound	Conc. (mg/m^3)	Compound	Conc. (mg/m^3)	Compound	Conc. (mg/m^3)
sulfur dioxide	260 ± 26	toluene	377 ± 38	formaldehyde	1.23 ± 12
nitrogen dioxide	190 ± 19	acetaldehyde	180 ± 18	ozone	1.96 ± 2
nitric oxide	123 ± 12	hexane	352 ± 35	ammonia	70 ± 7
hydrogen sulfide	140 ± 14	2-butanone	295 ± 30		
chlorine	290 ± 29	isobutanol	303 ± 30		
		dichloromethane	347 ± 35		
		tetrachloroethylene	537 ± 54		

MICRO-SCALE MODELING

Micro-scale modeling of the interaction of contaminants and sorbent media was carried out to better understand the processes and parameters involved, and ultimately to support macro-scale measurements and building modeling efforts. The main thrust of this work was to develop models that account for all the important processes thought to take place in an air cleaning device. In support of this, an attempt was made to characterize the pore system of the filter at the micrometer scale to supply a physical basis for permeability estimates of intra-bead permeability and diffusivity.

Transport Model

A numerical algorithm based on the lattice Boltzmann method was developed for modeling Navier-Stokes fluid flow in a bead pack geometry similar to that used in filter systems (Martys and Chen 1996; Nie and Martys 2007). The solution of flow fields obtained from the simulation is consistent with the Brinkman equation. That is, this model accounts for flow between beads as well as allowing for a non-zero bead permeability to account for possible flow through the beads. Once the local flow fields are determined, they serve as input into a code that solves for the advection, diffusion absorption and desorption of contaminants. The influence of flow through the bead on the absorption of contaminants in the bead pack can then be determined.

Filter Characterization

There are three levels of pore structure in a gaseous air-cleaning filter of the type considered in this modeling effort. The first is the pores between beads in the packing. This can be easily modeled, as will be seen below. The second is the pores in the beads that allow gas penetration into the beads. This pore scale was characterized via scanning electron microscopy (SEM) of cut beads and X-ray microcomputed tomography of whole beads, at the micrometer scale. The final level is the very small nanopores inside the beads that allow for sorption of contaminants. This level could potentially be characterized with transmission electron microscopy-based tomography and with focused ion beam tomography, but these techniques were not applied as part of the current project.

Figure 5 shows optical and SEM images of a sliced activated carbon-based bead. The optical images show the gross overall morphology of the beads, as well as the cross-section. SEM images show finer detail of the structure inside of a bead. Figure 6 shows one slice out of many slices making up an X-ray tomographic image of a bead. While the SEM pictures provide qualitative insight into the intra-bead pore system, a 3-D pore system model could not be derived from the limited micrometer-scale tomography due to low resolution of the micropores in the bead. X-ray tomography may be able to provide a more complete pore model in the future.

Figure 5: Optical and SEM images of beads. Top left – several beads, scale divisions are mm; top right – cross section of single bead; bottom left – SEM of cross-section, scale bar is 1 mm; bottom right – higher magnification image of cross-section, scale bar is 10 μm.

Model Test Case
Once the transport model was developed, the flow through a random packing of monosize beads was studied to determine the permeability of the total bead pack as a function of the permeability of the individual beads, since an experimentally-measured 3-D pore system model was not available. A monosize bead pack was used as an approximation, since the diameter of the actual beads did not vary by more than a factor of two. The beads were treated as continuum objects with a given permeability. The volume fraction of beads was about 0.50, which is typical for random deposition of mono size spheres. When modeling the dispersion, absorption and desorption of contaminant, it was assumed that the contaminant did not affect the actual flow field but moved passively. For test purposes, a simple linear absorption model was assumed.

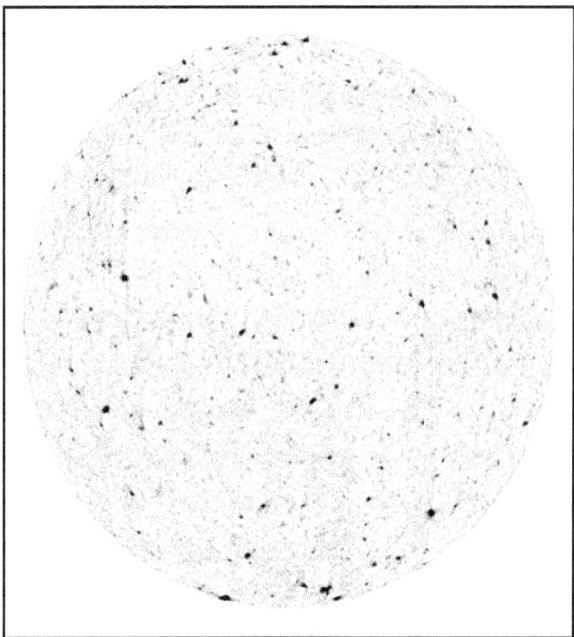

Figure 6: Cross-section slice from X-ray microcomputed tomography. Dark spots indicate pores. Scale is about 2.5 μm per pixel, and sample was about 4 mm across in this slice.

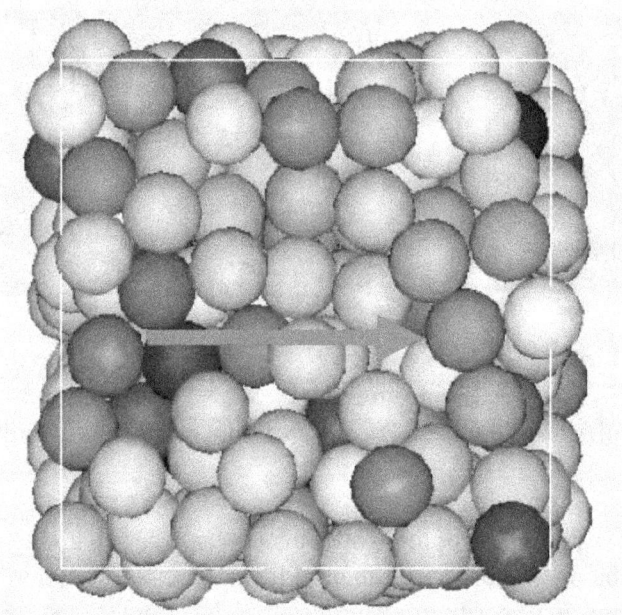

Figure 7: Bead pack, arrow indicates flow direction

Figure 7 illustrates a packing of spheres that was used in the simulation. The spheres may be inert solid or assigned a porosity, permeability, absorption/desorption coefficients and a relative diffusivity. The relative diffusivity is the diffusivity of a molecule relative to that in free space (air). That is, if a molecule has a certain diffusivity in air, its diffusivity is lower in the porous medium due to its interaction with the microstructure of the medium. These properties may be assigned individually, for each sphere, to allow for a distribution of properties as needed. A pressure drop is applied across the medium from left to right. Because periodic boundary conditions are maintained for the flow in the simulation, the fluid cycles through the medium. The system can be set up so that, initially, a concentration of contaminants is uniformly distributed. Alternately, a source or a sink term, allowing for addition or removal of contaminant at the inlet or outlet of the system, can be added.

Figure 8 shows the dependence of the total permeability of the filter on the permeability of the beads. For small values of the bead permeability, the vast majority of the flow goes between the beads and thus there is little effect on the overall permeability. As the bead permeability becomes 1/30th of the overall permeability, a sharp rise in the overall permeability begins. At very low bead permeability, the filter behaves like a packed hard sphere system. As the permeability of the beads is increased, two factors control the overall permeability of the system. Obviously the first cause of increase is the fact that there is additional flow though the beads. A second cause of increase is the fact that there is now an effective slip velocity at the bead surface corresponding to the original hard sphere surface. The slip velocity can significantly enhance total flow.

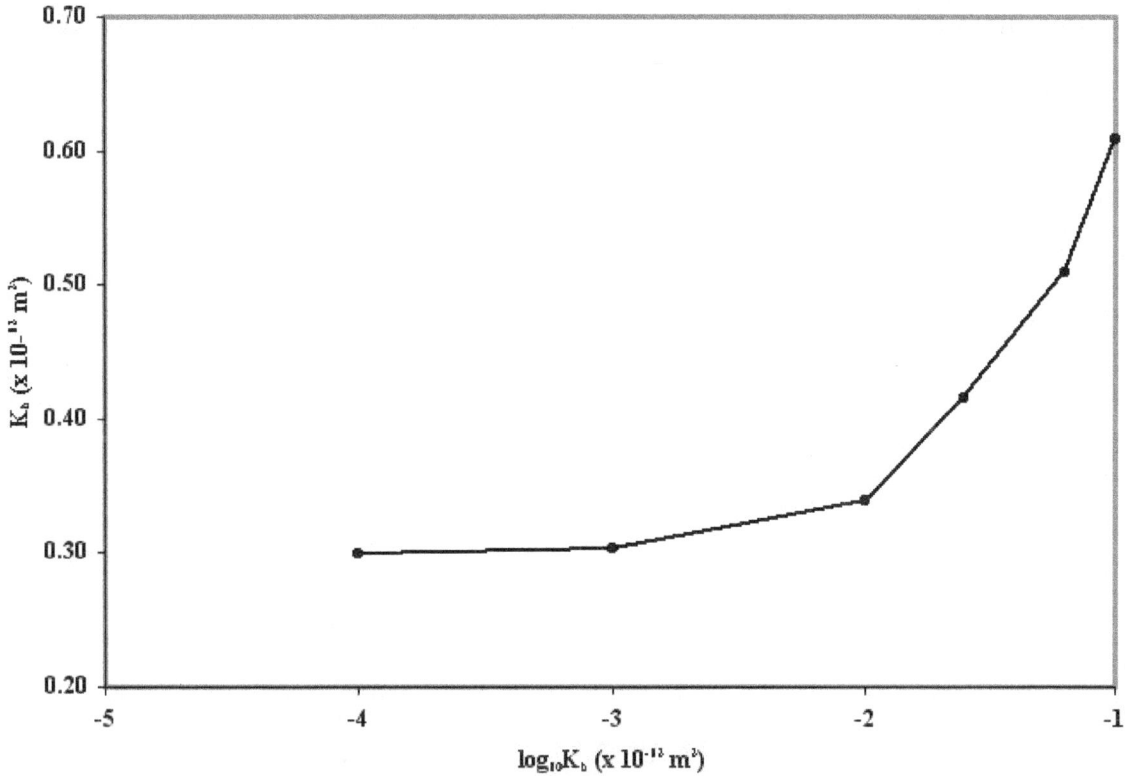

Figure 8: Bulk permeability of bead pack as a function of the individual bead permeability. The arrow indicates the bulk permeability in the limit of zero bead permeability.

Modeling contaminant transport
The transport of contaminants in the fluid phase is described by the advection-diffusion equation $\frac{dC(t)}{dt} = -\vec{\nabla} \cdot C\vec{V} + D\nabla^2 C$ where D is the molecular diffusion coefficient of the contaminant in the fluid. A dimensionless number, Peclet number, $Pe = Vl/D$, where l, is a typical length scale, taken to be the bead diameter, that characterizes the relative importance of flow vs. diffusion in

transport phenomena. Here cases where *Pe = 1* and *10* are considered for the flow fields determined in the simulation. These values of Peclet number are representative of flow rates through filters corresponding to the lower flow rate regime. As the contaminant is transported, it may be adsorbed into or desorbed from the beads. This phenomena is modeled by including a sink/source term to the transport equation $\frac{dC_s(t)}{dt} = \alpha C(t) - \beta C_s(t)$. This is a simple linear model of the process, used for convenience, where C_s corresponds to the concentration in the porous medium, and α and β are adsorption and desorption coefficients. Figures 9 and 10 illustrate how the performance of a filter system may be studied with this numerical approach. Figure 10 is the same case as that in Fig. 9, but with the diffusivity of the contaminant in a bead reduced by a factor of ten. All other factors, bead packing, initial concentrations, flow rates, and adsorption/desorption coefficients were kept the same. In this flow regime *Pe = 10*, so that hydrodynamic forces are more important than the diffusive motion of the contaminant. The concentrations of contaminant in the macro-pore space (blue), in the bead micropores (red), and in absorbed contaminants (green) were tracked in time. Clearly the efficiency of the filter in removing contaminants diminishes as the diffusivity of the contaminant in the bead decreases.

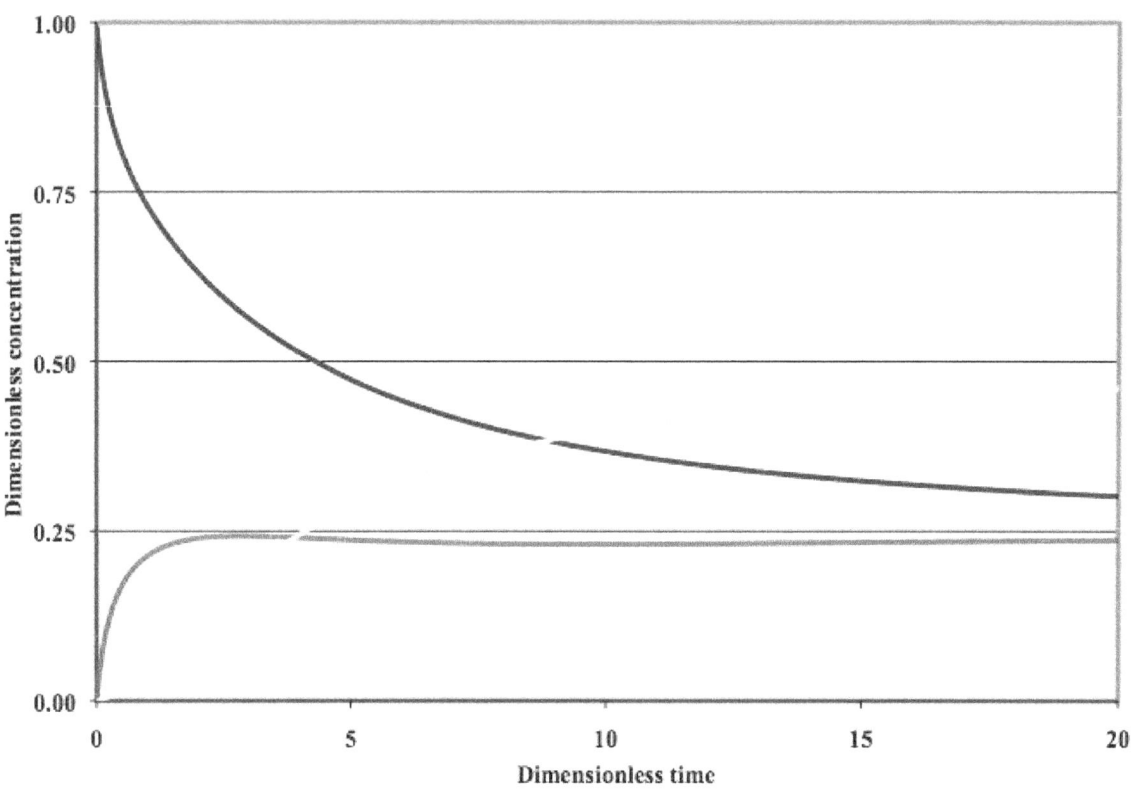

Figure 9: Relative concentration vs. time in filter. The blue line is the concentration of contaminant in the macro-pores. The red line is the contaminant in the bead micropore space and the green is that adsorbed into the bead material. Note, the units of concentration here is dimensionless as we normalized the concentration of contaminant found in either the micropores or that absorbed by the beads by the total contaminant initially found in the macro-pores space. This way we know the fraction of contaminant residing in each part of the microstructure.

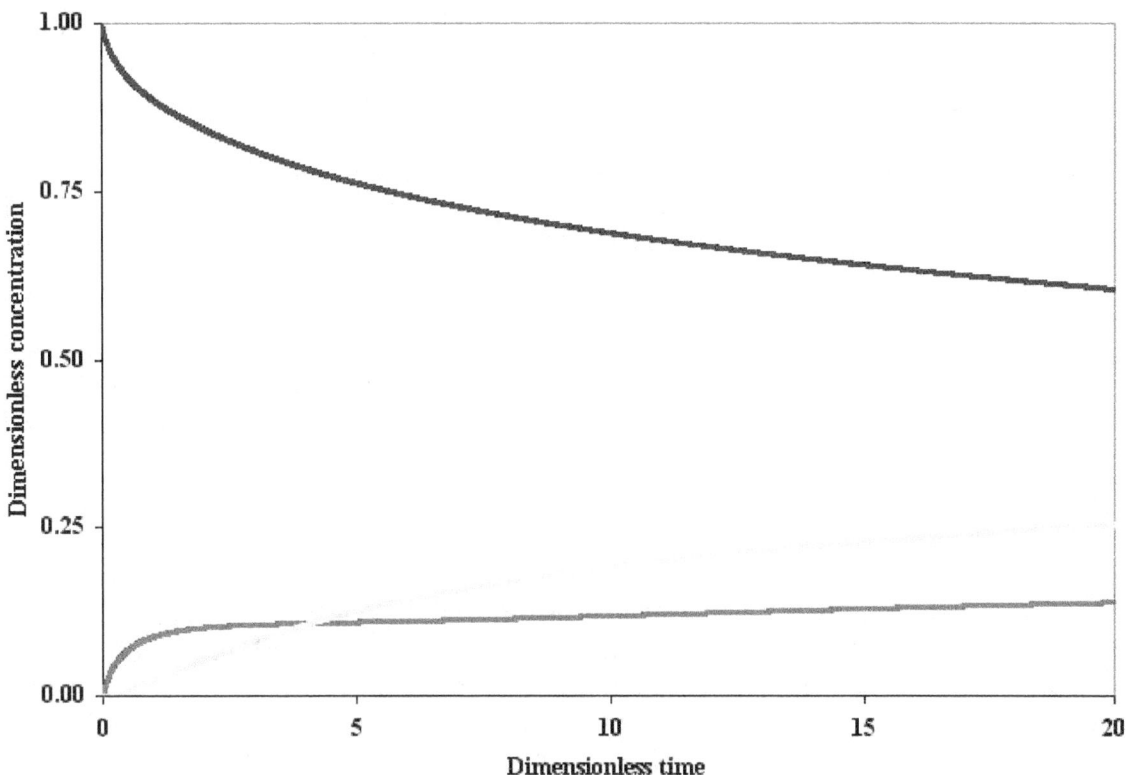

Figure 10: Relative concentration vs. time in filter. Same color scheme as Fig. 9. In this case, the diffusivity of the contaminant was reduced by a factor of 10. As a result, the capability of rapid absorption is significantly reduced.

Link to measurements of concentration over time
It is important to note that the model used in this study is phenomenological and needs to be validated by experiment or replaced by a model based on experimental data for the medium concerned. This section describes one way in which this goal may be attained. In the case of the linear model used in this study, the values of absorption/desorption coefficients may be obtained from study of the concentration vs. time profiles associated with a dispersed contaminant in contact with a material (e.g. media used in filter) of interest. In general, depending on the form of the concentration vs. time profile, suitable reaction rate equations can be identified whose coefficients can be fit to the data. For example, let us assume that the time value concentration profile is well described by an exponential decay to some equilibrium. That is,
$C(t) = (C_0 - C_\infty) \exp(-kt) + C_\infty$, where $C(t)$ is the concentration, in units of mass/volume, as a function of time, t, C_0 is the initial concentration, C_∞ is the final or equilibrium concentration and the decay rate, k, is determined by fitting to the data. Such profiles can be related to a linear reaction rate model describing the rate of adsorption of contaminant on a surface
$\frac{dC_s(t)}{dt} = \alpha C(t) - \beta C_s(t)$, where $C_s(t)$ is the amount of contaminant (mass) per surface area, and α and β are rate coefficients. For this model, it can be shown that $\alpha = k\frac{V}{S}(1 - \frac{C_\infty}{C_0})$ and

$\beta = k \dfrac{C_\infty}{C_0}$, where V is the contaminated fluid volume in the experimental apparatus and S is the bead surface area. The above results assume that the concentration of contaminant corresponds to a dilute regime, the contaminant is uniformly mixed in the experimental cell and the adsorption process is reversible. Including effects such as irreversible reactions may be possible to model by including a sink term in the surface interaction term. Such analysis may also be extended to other cases such as when there is an upper bound to the amount adsorbed by a surface

Future Research Needs
This section described a model of the flow between and in microporous materials. It was demonstrated that flow in small micropores is highly restricted once the bead permeability was below some value. It was also found that the absorption is greatly diminished in this case. The effects of changing bead diffusivity were also modeled and the effect on the concentration of contaminant in the macro-pore space between the beads, in the bead micropores, and absorbed contaminants inside the beads were demonstrated.

Future research needs include a validation program to compare simulation results with experiments. While accurate flow and contaminant transport simulations are possible, to predict the degree of contamination of a bead filter system, a boundary condition is needed that describes the rate at which the contaminant is adsorbed or desorbed from a surface. This may be obtained by study of the concentration vs. time profiles in experiments similar to that used for determining isotherms. As there are a plethora of phenomenological equations describing reaction rates, it is important to link parameters of models with experimental data when possible. Indeed, for a computational model to be a successful predictor, such accurate materials characterization is crucial. Increased microstructure characterization efforts, including focused ion beam and transmission electron microscopy-based tomography should give interesting and useful results and provide a more physical basis for bead properties.

BUILDING SCALE MODELING
In order to assess the potential impacts of gas-phase air cleaning on building occupant exposure, whole building simulations were performed using network modeling. These simulations are based on a previous study of retrofits for improved building protection (Persily et al. 2007), however the simulations reported here are more detailed. The building simulations employed the multizone airflow and contaminant dispersal model CONTAM (Walton and Dols 2005), which was applied to three generic buildings intended to represent U.S. commercial buildings.

The three buildings include a single-zone one-story building, a two-story building with more realistic zoning, and a high-rise office building with central air handlers. Details on the buildings are provided in Persily et al. (2007), including building layout, occupancy levels, and ventilation systems. Table 4 describes the size and system types of the three buildings. The systems are assumed to have constant airflow rates, with the outdoor air intake rate equal to 10 % to 20 % (depending on the building) of an assumed supply airflow rate per unit floor area of 5 L/s•m^2 (roughly 1 cfm/ft^2). The envelope infiltration rate in the single-zone building is assumed to be constant at an air change rate of 0.2 h^{-1}; in the other two buildings the infiltration rate is calculated with CONTAM based on airtightness values expressed as effective leakage areas at 4 Pa of 5 cm^2/m^2 of wall area (0.07 in^2/ft^2) in the two-story office and 8.7 cm^2/m^2 (0.13 in^2/ft^2) in the high-rise. The baseline building models have no gas phase air cleaning.

	# of stories	Floor area, m² (ft²)	Ventilation System Type
1, Single-zone	1	1000 (10 800)	Simple air handler
2, Office	2	2600 (28 000)	Single air handling system
3, Office	14	11,900 (128 000)	Central air handling systems

Table 4 Simulated Buildings

The simulations employed a nonreactive gaseous contaminant and the following four release locations:

    Exterior release distant from building
    Exterior release at outdoor air intake(s)
    Interior release in lobby
    Interior release into ventilation system return

The outdoor, distant release is represented by a constant, elevated outdoor concentration of 1 mg/m³ for a period of 60 s. The release at the intake is modeled as a localized increase in the outdoor concentration at that specific location, again lasting 60 s. The indoor releases are represented by contaminant release rates of 10 g/s for 60 s in the designated location. Only the outdoor release was considered for the single-zone building.

The calculated concentrations and the assumed release rates have no significance in relation to any particular chemical agent but were chosen to yield indoor concentrations in a reasonable range of interest. Given the generic nature of these contaminants and releases, and in keeping with the purpose of the project, the calculated concentrations cannot be used to estimate health impacts. As a result, the simulation results in terms of the relative concentration or exposure between the various simulated cases are of far more relevance than the absolute concentration and exposure values themselves.

The air cleaning options considered in the simulations include removal efficiencies of 95 %, 97.5 %, 99 %, 99.9 % and 99.99 %, located in either the outdoor air intake or the mixed airstream (downstream of where the outdoor air intake and recirculation air mix). In addition, a tighter envelope, corresponding to an air change rate of 0.01 h^{-1} in the single zone building and effective leakage areas of 0.7 cm²/m² (0.01 in²/ft²) in the two-story and high-rise buildings, was combined with the gaseous air cleaner in the outdoor air intake. Also in these two buildings, the increased envelope airtightness and enhanced outdoor air filtration were combined with a doubling of the outdoor air intake rate in an effort to pressurize the building interior.

These simulations were run for 12 h with a 5 s time step with a wind speed of 5 m/s (11 mi/h) and an indoor-outdoor temperature difference of 20 °C (36 °F). The simulations yield contaminant concentrations as a function of time in each building zone, which were subsequently used to determine occupant exposures. The exposure of an individual building occupant is the average contaminant concentration to which he or she is exposed over the simulation period in units of mg·min/m³. The average exposure for all occupants of the building is then determined from the individual occupant exposures. These exposures are calculated over 6 h, starting 1 h before the release and continuing 5 h after. For each simulation, the average building exposure is compared with the baseline exposure, i.e., without any gaseous air cleaning, to determine the exposure as a percentage of the baseline exposure. Therefore the baseline case corresponds to 100 %, while a reduction in exposure results in a value below 100 %.

Simulation Results

Figure 11 is a plot of the simulated concentrations for the single zone baseline and enhanced filtration cases with a 95 % removal efficiency, subject to the outdoor release. The baseline case is the solid line with a peak concentration just under 0.025 mg/m^3. The other lines show the impacts of air cleaner location (outdoor or mixed airstream) and of combining reduced infiltration (airtightening) with the air cleaner located in the outdoor air intake. Locating the air cleaner in the outdoor air intake reduces the indoor concentrations significantly, but the mixed air location is more effective given the high recirculation airflow rates. Tightening the building results in significantly less contaminant entering the building due to infiltration, corresponding to the lowest concentrations in Figure 11.

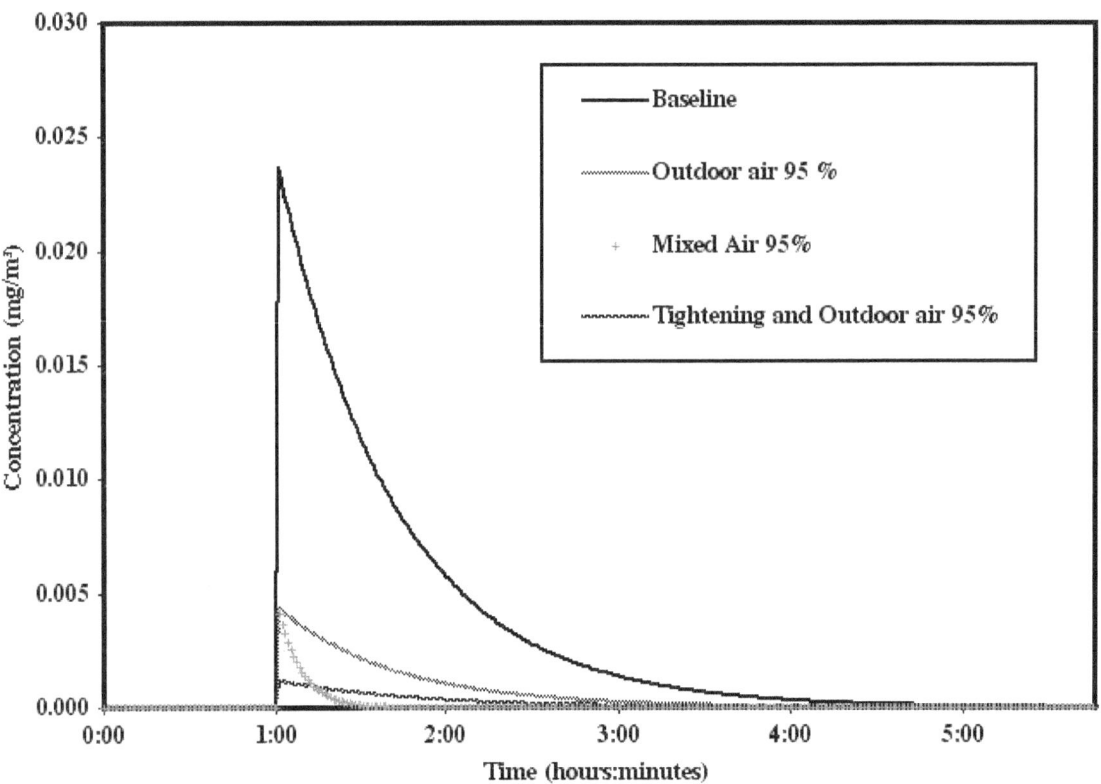

Figure 11 Simulation results for single-zone model with 95 % efficient air cleaner

Figure 12 shows the simulation results for the outdoor air intake release over the five values of the removal efficiency. The incremental concentration reductions associated with the increased efficiencies are evident, but the existence of a significant infiltration rate results in diminishing returns given that this infiltrating air is not cleaned.

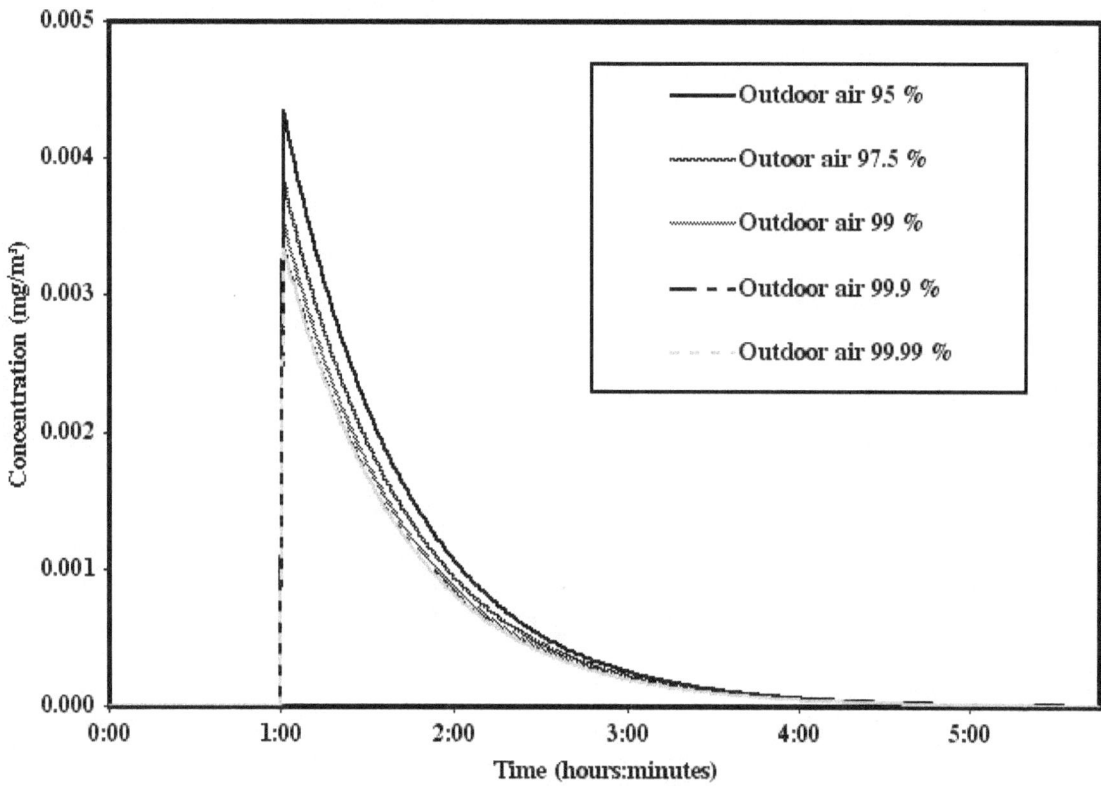

Figure 12 Simulation results for single-zone model with varying removal efficiencies

Figure 13 summarizes all the simulation results for the single-zone model in terms of the exposure relative to the baseline case with no gas phase air cleaning. This figure shows the higher impact of locating the air cleaner in the mixed air relative to the outdoor air, as well as the exposure reductions associated with airtightening. However, the residual infiltration for each air cleaning configuration is seen to limit the exposure reductions that can be achieved with higher removal efficiencies.

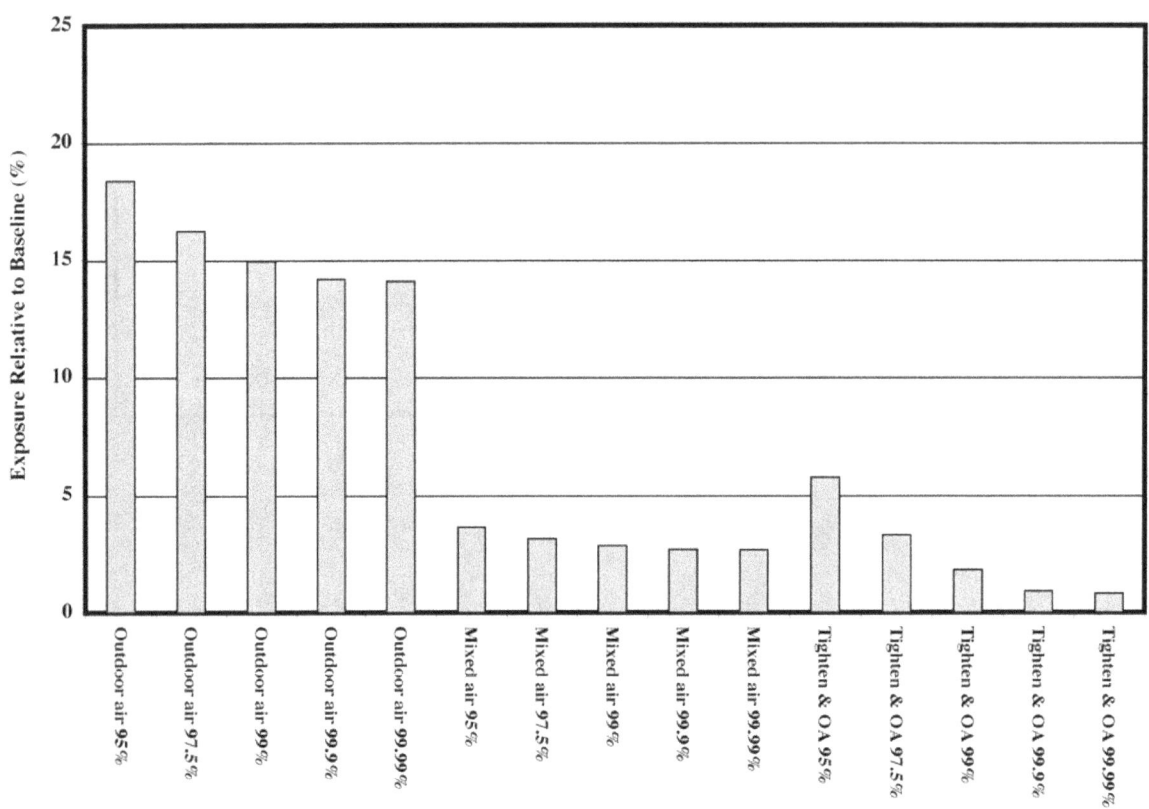

Figure 13 Summary of the simulation results for single-zone model

Figures 14 and 15 summarize the simulation results for the two-story and high-rise office buildings. In these cases, four release scenarios were considered: a general increase in the outdoor concentration, a local release at the intake, and releases in the building lobby and a return vent. The two indoor releases were only considered for the mixed-air cases. As was the case for the single-zone simulations, the mixed air location is seen to be more effective than cleaning only the outdoor air. However, for both locations, the residual infiltration limits the overall reduction that is achievable through increased removal efficiencies. Reducing infiltration by tightening the building or increasing the outdoor air intake in an attempt to pressurize the building results in very low relative exposures at the higher efficiencies.

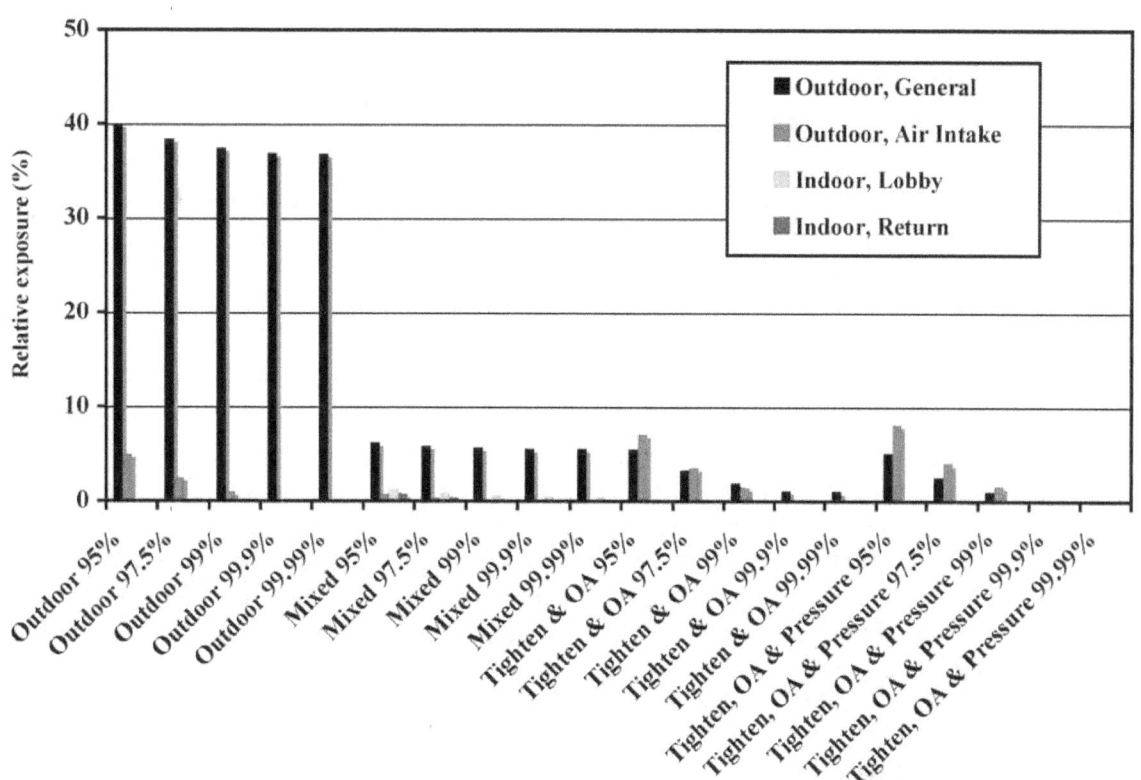

Figure 14 Summary of the simulation results for two-story office building

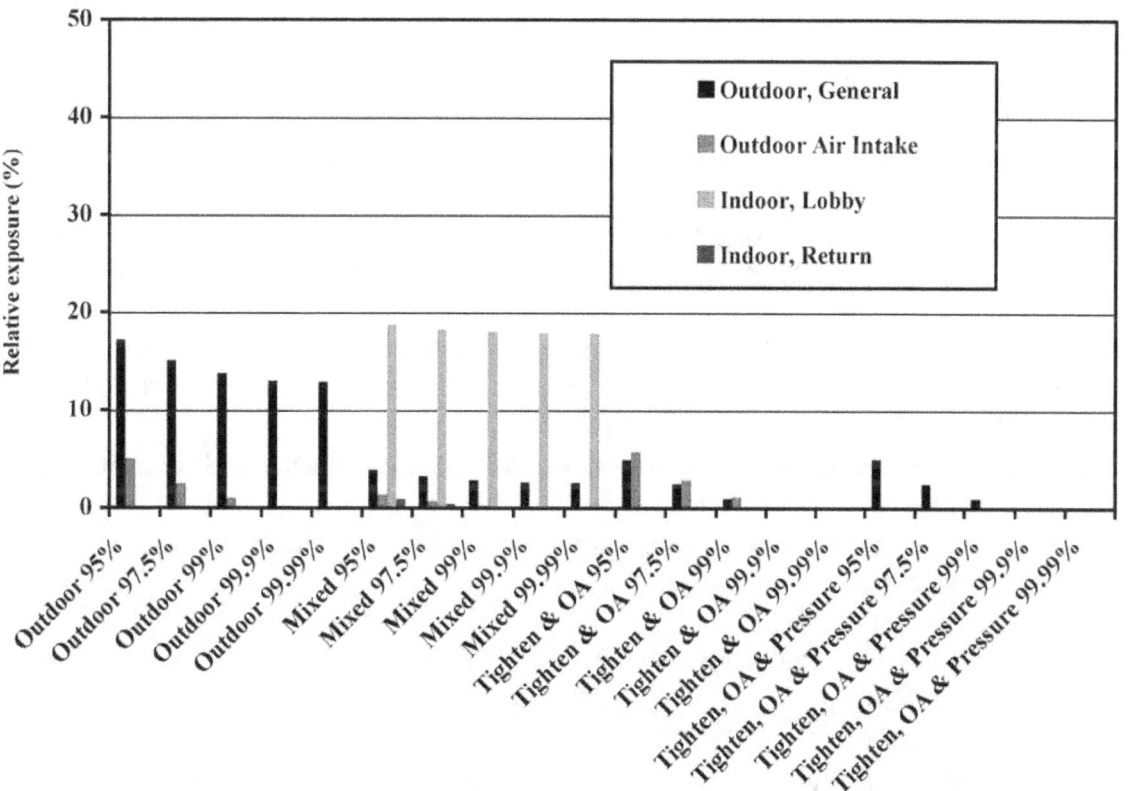

Figure 15 Summary of the simulation results for two-story office building

The whole building simulations show the significant reductions in exposure that can be achieved through increasing levels of gas-phase air cleaning. However, they also show the importance of system location and the decrease in overall effectiveness that will occur when outdoor contaminants enter a building through infiltration, which does not pass through the air cleaning system. Reducing such infiltration through building tightening and increased levels of outdoor air intake will increase the overall air cleaning effectiveness.

DISCUSSION

Gas phase air cleaning is playing a key role in building protection strategies, which is likely to increase as new technologies are developed and new design and performance criteria are developed. In order to meet these new demands, standardized methods of test for such systems need to be developed to support design decisions along with standards and other guidance on installation, operation and maintenance activities. While some such standards have been progressing, such as ASHRAE Standard 145.1P, these needs still exist. Additional research is needed to define more reliable and perhaps cheaper and faster laboratory test methods and to relate the results of these tests to installed performance in the field. The efforts described in this report represent some initial steps in meeting these needs, but additional work needs to be pursued to support the standards that are required into the future.

ACKNOWLEDGEMENTS

This work was supported by the Directorate of Science and Technology of the U.S. Department of Homeland Security,

REFERENCES

Alberici, R.M., Jardim, W.F. 1997. Photocatalytic destruction of VOCs in the gas-phase using titanium dioxide. *Applied Catalysis B: Environmental*, **14**: 55 - 68.

ASHRAE. 2004. Ventilation and Acceptable Indoor Air Quality, ASHRAE 62.1-2004. American Society of Heating, Refrigerating, and Air-Conditioning Engineers.

ASHRAE. 1999. Method of Testing General Ventilation Air-Cleaning Devices for Removal Efficiency by Particle Size, ASHRAE 52.2-1999. American Society of Heating, Refrigerating, and Air-Conditioning Engineers.

ASHRAE. 1992. Gravimetric and Dust Spot Procedures for Testing Air-Cleaning Devices Used in General Ventilation for Removing Particulate Matter, ASHRAE 52.1-1992. American Society of Heating, Refrigerating, and Air-Conditioning Engineers.

ASME. 2002. Nuclear Power Plant Air-Cleaning Unit and Components, ASME N509-2002. American Society of Mechanical Engineers.

ASME. 1989. Testing of Nuclear Air Treatment Systems, ASME N510-1989. American Society of Mechanical Engineers.

ASTM. 2004. Standard Test Method for Nuclear-Grade Activated Carbon, ASTM D3803-91(2004). American Society of Testing Materials.

Blake, D.M. 1994. Bibliography of Work on the Photocatalytic Removal of Hazardous Compounds from Water and Air, NREL/TP-430-6084. National Renewable Energy Laboratory.

Boeniger, M.F. 1995. Use of Ozone Generating Devices to Improve Indoor Air Quality. *American Industrial Hygiene Association Journal*, 56: 590 – 598.

Chen W, Zhang JS, Zhang Z. 2005. Performance of Air Cleaners for Removing Multiple VOCs in Indoor Air. ASHRAE Transactions; 112 (I): 1101 - 1114.

Daniels, S.L. 2002. On the Ionization of Air for Removal of Noxious Effluvia. IEEE Transactions on Plasma Science, 30 (4): 1471 – 1481.

Godish, T. 1989. *Indoor Air Pollution Control*. Lewis Publishers, Inc., Chelsea, Michigan.

Henschel, D.B. 1998. Cost Analysis of Activated Carbon Versus Photocatalytic Oxidation for Removing Organic Compounds from Indoor Air. *Journal of the Air & Waste Management Association*, 48: 985 – 994.

Jacoby, W.A., Blake, D.M., Fennell, J.A., Boulter, J.E., LeAnn M. Vargo, George, M.C., Dolberg, S.K. 1996. Heterogeneous Photocatalysis for Control of Volatile Organic Compounds in Indoor Air. *Journal of the Air & Waste Management Association*, 46: 891 - 898.

IEST. 2005. Testing HEPA and ULPA Filter Media, No. 0-9747313-3-1. Institute of Environmental Sciences and Technology, Contamination Control Division Recommended Practice 021.2.

IEST. 2005. HEPA and ULPA Filter Leak Tests, No. 0-9747313-4-X. Institute of Environmental Sciences and Technology, Contamination Control Division Recommended Practice 034.2.

IEST. 2004. Testing Cleanrooms, No. 1-877862-99-1. Institute of Environmental Sciences and Technology, Contamination Control Division Recommended Practice 006.3.

IEST. 1995. A Glossary of Terms and Definitions Relating to Contamination Control, IEST-RD-CC011.2. Institute of Environmental Sciences and Technology, Contamination Control Division Reference Document 011.2.

IEST. 1993. HEPA and ULPA Filters, IEST-RP-CC001.3. Institute of Environmental Sciences and Technology, Contamination Control Division Recommended Practice 001.3.

IEST. 1992. Testing ULPA Filters, IEST-RP-CC007.1. Institute of Environmental Sciences and Technology, Contamination Control Division Recommended Practice 007.1.

Martys, N.S., Chen, H. 1996. Simulation of multicomponent fluids in complex three-dimensional geometries by the lattice Boltzmann method. *Phys. Rev. E*, 53(5), 743-750.

Nie, X., Martys, N.S. 2007. Breakdown of Chapman-Enskog expansion and the anisotropic effect for lattice-Boltzmann models of porous flow, *Physics of Fluids*, 19 (1), 011702-011702-4.

NAFA. 1997. Installation, Operation and Maintenance of Air Filtration Systems. National Air Filtration Association.

NAFA. 2001. NAFA Guide to Air Filtration, Third Edition. National Air Filtration Association.

NIOSH. 2003. Guidance for Filtration and Air-Cleaning Systems to Protect Building Environments from Airborne Chemical, Biological, or Radiological Attacks, Publication No. 2003-136. Department of Health and Human Services, National Institute for Occupational Safety and Health.

Persily, A.K., Chapman, R.E., Emmerich, S.J., Dols, W.S., Davis, H., Lavappa, P. and Rushing, A. 2007. Building Retrofits for Increased Protection Against Airborne Chemical and Biological Releases. NISTIR 7379, National Institute of Standards and Technology.

VanOsdell, D.W., Owen, M.K., Jaffe, L.B., Sparks, L.E. 1996. VOC Removal at Low Contaminant Concentrations Using Granular Activated Carbon. Journal of the Air & Waste Management Association, 46: 883-890.

Walton, G.N. and Dols, W.S. 2005. CONTAMW 2.4 User Guide and Program Documentation, NISTIR 7251, National Institute of Standards and Technology.

Weschler, C.J. 2000. Ozone in Indoor Environments: Concentration and Chemistry. Indoor Air *2000*, pp. 269-288.

www.ingramcontent.com/pod-product-compliance
Lightning Source LLC
Chambersburg PA
CBHW081813170526
45167CB00008B/3418